GREENING YOUR OFFICE

Greening
Your Office

From cupboard to corporation: an A–Z guide

Jon Clift & Amanda Cuthbert

Green Books

Published in 2007
by Green Books, Foxhole, Dartington, Totnes, Devon TQ9 6EB

Design concept by Julie Martin jmartin1@btinternet.com

Printed by Cambrian Printers, Aberystwyth, UK on Revive 50, made from 50% recycled material

DISCLAIMER: The advice in this book is believed to be correct at the time of printing, but the authors and publishers accept no liability for actions inspired by this book.

ISBN 978 1 900322 14 0

We would like to thank Ben Brakes of Barclaycard, Simon Clifford of BBC Somerset Sound, the Britannia Building Society, Rory McMullan, author of *Cycling to Work: a beginner's guide,* Paul Ellis of the Ecology Building Society, Amy Sims of Global Action Plan, Ali Clabburn of Liftshare, Matt Criddle of NatureSave, Chris Hillyer of South Hams District Council and Heather Goringe of Wiggly Wigglers.

> *To discuss a possible bulk purchase of this book, please phone Green Books on 01803 863260*

Contents

Introduction

"In my view, climate change is the most severe problem we are facing today, more serious even than the threat of terrorism."
– Professor Sir David King, Chief Scientific Advisor to H.M. Government

Climate change is happening, due primarily to rising carbon dioxide (CO_2) levels in the atmosphere; we are running out of landfill sites, and using up our natural resources at an alarming rate. Whether you are a managing director or a receptionist, a global corporation or a one-man band, there are things you can do – both as an individual and as an organisation – to reduce your carbon footprint and help both the planet and your budget. In writing this book we have concentrated on the administrative side of a business, looking at what can be done to make the office a greener place; there is not space in a book of this kind to address all the things which companies as a whole can do to green up, since they come in many different shapes and sizes and have differing concerns, depending on their nature. However, many of the principles, such as saving energy and water, can be applied to other areas of a business as well as the office.

We have included inspirational examples of those who have greened up and benefited from doing so. If you make just one office practice more environmentally friendly, you will make a difference.

Chapter 1

Why do it?

We spend a large amount of our time in the office, where we use considerable amounts of energy and resources. 'Greening the office' will therefore have significant benefits, both environmentally and financially. As many other businesses have experienced, small changes to your office practices and procurements, mostly costing nothing, will reduce the company's expenditure and improve your sales. By being proactive, you stay ahead of impending government legislation, as climate change and mandatory reductions in carbon emissions are being debated. It makes compelling, realistic business sense to green up the office, and helps in our battle with climate change.

"For every £1 invested now we can save £5, or possibly more, by acting now." – Sir Nicholas Stern, *The Stern Review*, 2007

There are many arguments for going green in the office:

- Minimise overheads by reducing
  - use of consumables
  - energy costs
  - costs of waste disposal
  - water bills
  - expenditures on office hardware
  - transport bills
  - parking spaces

- Increase competitiveness by:

  - keeping production costs down as overheads are cut
  - improving sales as your green credentials attract new customers
- Improve investment potential: increasing numbers of investors only invest in businesses that have environmentally responsible policies
- Create a feel-good factor in the workplace, encouraging staff retention and recruitment
- Enhance your reputation and brand awareness
- Do your bit to reduce climate change

SOUND SENSE

Simon Clifford, Editor of BBC Somerset Sound, led the way to green up the organisation – as a radio station they launched a Somerset-wide campaign to green up the county, but wanted to get their own house in order first. They started by measuring their carbon footprint to get a baseline for improving their own performance, and then put into motion some recommendations. They swapped pool cars for more fuel-efficient car-derived vans; converted their fleet to a 15% biofuels product; banned personal cars in the car park and ordered a flexi-fuel car. They improved insulation in the building and increased their recycling of paper, food, glass, tin etc; cancelled their water-cooler contract; swapped their free-supplied coffee/tea making machine in favour of paid-for fair trade alternatives, and opted out of the national BBC cleaning contract in favour of green alternatives. They no longer give away leaflets, but instead are using more innovative things like reusable memory sticks to get their local messages over. They are paying to offset CO_2 emissions.

The reaction amongst the staff was extremely positive – when the offsetting figure was arrived at, 15 out of 16 of the staff voluntarily agreed to pay from their own pockets rather than use licence-payers' money. They now have regular carbon footprint audits, and one of their reporters monitors their progress – e.g. switching off computers, chargers etc – and is also responsible for driving the campaign editorially.

www.bbc.co.uk/somerset/local_radio

Chapter 2

Getting started

"On average an office wastes £6,000 each year by leaving equipment on over weekends and bank holidays." – Carbon Trust

Whether you are part of a large organisation or a one-man band, this book will help you green your office. Even if you do only some of the things suggested below, both your business and the environment will benefit.

There are various organisations available to guide you, with financial assistance and grants available for many projects. Many other offices have already successfully implemented a green agenda, and people are willing to help you to do the same by sharing their best practices.

Even when you are merely contemplating greening your office, there is generally help available, depending upon where you are and the size of your organisation. Organisations such as Envirowise offer initial confidential visits, and should you wish to proceed, a support package tailored specifically for your business. They will also be able to advise you on grants and other assistance available within your region, and provide opportunities to meet and discuss ideas and issues with other businesses at Resource Efficiency Clubs. **www.envirowise.gov.uk**

For small to medium-sized companies, charities such as Global Action Plan offer support and advice specifically for your needs. They have an impressive track record, but only operate in some parts of the country. There are grants and subsidies available in many cases, both for using Global Action Plan and for the implementation of recommended changes. **www.globalactionplan.org.uk**

PAPER FIGURES

Through Global Action Plan's Environment Champions programme, Britannia Building Society recruited a team of enthusiastic employees who volunteered to measure resource consumption and waste generation in their places of work and to encourage their colleagues to help to reduce these impacts.

These Champions initially measured the environmental impacts of the two head office sites in Leek, and the offices in Plymouth and London. After seeing the problems at first hand, the volunteers came up with fitting actions that their colleagues could take to improve paper use, energy consumption, CO2 emissions, waste generation and recycling.

With the project backed by the management of the company, the Champions then set about communicating these simple actions to the whole workforce. Their campaign designated three individual months in autumn 2005 to energy use, paper consumption and recycling. During these months, the Champions ran high-profile events. In Energy Use month, the group built an exercise bike that generates electricity to highlight energy efficiency messages, and placed good-natured reminders on people's screens if they left their monitors on overnight. In Paper month, the group created a sculpture of empty paper boxes in the office reception that represented the amount of paper used in a few hours. In Recycling month, the TV character "Dusty Bin" was brought to life to walk the offices of Britannia, communicating the 3 R's of reduce, reuse and recycle, and a waste quiz was run for employees to participate in.

The Champions re-audited their offices after the campaign, and discovered that their efforts had achieved the following changes:

- Leek – half-yearly paper consumption had fallen by 2.25 million sheets, which would stand over twice the height of Big Ben
- London – paper recycling and plastic recycling increased by 59% and 32% respectively
- Plymouth – total waste generated had decreased by 5%
- Leek – paper sent to be recycled had increased by 11%
- Plymouth and London – paper content of the waste sent to landfill had decreased by 46% and 59% respectively
- Leek – emissions from energy use had fallen by almost 3 million footballs-worth of CO_2. This avoided over £3,000 in energy bills in the first months after their campaign.
- London – monitors left on overnight reduced by 44%
- Plymouth – monitors and PCs left on out of hours were reduced from 40% to 5% and 17% respectively.

A survey of staff attitudes at the Leek head office showed that 73% of employees had changed their habits at work due to the work of the Champions, and 30% had changed their habits at home.

As well as engaging their colleagues in behaviour change, the Champions successfully lobbied their facilities management department for improved recycling facilities and duplex printing. As a result they were able to save over 300,000 sheets of paper in 2006.

Global Action Plan is the practical environmental charity that helps people to make positive changes at home, at work, at school and in the wider community.

www.globalactionplan.org.uk www.britannia.co.uk

Chapter 3

Getting people on board

Staff are key to the whole process of greening the office. They are an incredible resource, and if asked, will come up with numerous thoughts and suggestions to minimise waste and reduce energy costs.

From the very beginning they need to be part of the process, involved and informed. It is obviously crucial that senior management are also committed to the process – that they 'walk the talk' and lead the way.

With climate change now a factor of life, most people will be keen to help green up the office, especially when they are encouraged to own the process.

Motivate staff and colleagues

If your organisation is spread across several sites, you can start small in one of them and spread the scheme to the others. Depending on the size of the organisation and whether all staff work in one place, there are a number of ways to get the ball rolling:

- A staff suggestion box
- An internal email address
- A 'green' office board
- An intranet forum discussion group
- An initial meeting if feasible

WHERE CREDIT'S DUE

The impetus for the greening-up process at Barclaycard came from the staff, who wanted to know what the company was doing to reduce CO_2 emissions. As a result Barclays Group employed a dedicated Environmental Manager, Ben Brakes, who set up 5 year targets. They started with recycling, gradually improving the scheme over the years until in 2006 the UK group were awarded the Green Apple Award for best practice in the Finance and Banking category.

They then moved on to energy, and ran energy awareness campaigns such as the "Switch it off" message, designed to have monitors switched off over night and when not in use. Staff response to the initiatives was excellent – recycling rates are constantly high, and they are always looking at expanding the scheme; a staff survey showed that staff felt Barclaycard was a responsible employer. Barclaycard also introduced a car-share scheme, and as a result they now have around 20% of their staff car-sharing.

Ben has introduced a network of Environmental Champions made up of committed staff members who can promote the environmental initiatives taking place and feedback successes and suggestions to him. He says "The main gain from a lot of this is good PR. It is vital now for any business to be successful to be seen to be, not only 'green', but really pushing the agenda forward."

In early 2006 Barclaycard, along with several other sites in the Barclays Group, achieved ISO 14001 accreditation for its Environmental Management System. In 2007 Barclaycard, as part of Barclays, went carbon-neutral for all its UK operations.

www.barclaycard.co.uk

Create an 'Environmental Taskforce'

Ask for volunteers to set up an 'Environmental Taskforce' to involve staff, collate suggestions and implement realistic ideas. If your organisation is large, you will need a co-ordinator and individuals in each section. If you have a smaller office this could be done by one person.

The Environmental taskforce will:

- Monitor current practices: for example, sheets of paper used per week, number of staff travelling by car, number of computers left on stand-by each night, number of lights left on in the office, etc

- Set realistic targets. SMART targets are great to keep in mind: **S**pecific, **M**easurable, **A**ttainable, **R**ealistic, **T**ime-bound

- Start with 'easy hit' targets (e.g. reducing paper consumption, recycling ink cartridges, switching off monitors, switching lights off when not needed)

- Consider appropriate incentives: a carrot is better than a stick!

- Nominate individuals within the team to monitor progress on each project being tackled

- Keep the communication flowing. It's crucial to keep everybody well informed and up-to-date regarding targets set, achievements, and any problems encountered

- Let the outside world know what's happening. Use the process to everyone's advantage; use in-house journals, the press, your website, and local TV and radio – and tell your customers. Greening up is great PR

- Celebrate success. Think up intriguing and innovate ways to mark milestones

- Keep the momentum going. Once 'easy hit' targets have been achieved and the workforce is motivated and 'on-board', try for more ambitious (but still SMART) targets

- Beware of a false sense of completion – whilst a few simple 'quick wins' are great for morale, they can and frequently do make people think that the job is done

Every individual act makes a difference. Whether you work from home or are part of a global corporation, your actions big and small will contribute to the fight against climate change.

GREEN FOUNDATIONS

When the Ecology Building Society was founded in 1981, it was for the sole purpose of providing ecological mortgages and promoting sustainable living. With their lending policy being so environmentally focussed, it was only natural that their office and business practices followed suit. Initially they employed local people who could walk to work, used recycled paper exclusively and minimised waste. Now they operate from a purpose-built eco-office, and have a comprehensive environmental policy which includes generating some of their own electricity, composting their biodegradable waste and recycling rainwater.

Ecology's staff have always been quite assiduous in complying with agreed practice, recognising a duty of care to the Society's members and their expectations. Staff diligently turn off appliances where possible, empty the composting bin and bring their used batteries from home so that they can be aggregated for taking to the recycling centre.

Every member of staff is involved in maintaining the Society's green credentials and contributing ideas to reduce its environmental impact. The Chief Executive undertakes monitoring. Some of their initiatives have lead to direct reductions in running costs – the Society's water bills are minimal.

www.ecologybuildingsociety.co.uk

A-Z Guide

Accounts – *see Paper*

Adhesives – *see Glues, Solvents*

Aerosols

Where possible avoid aerosols: both the propellant and the contents could damage your health. Buy environmentally-friendly products that come in liquid form, and wipe or spray using a pump spray.

Air Conditioning – *see also Lighting, Office Equipment, Plants*

Running an air-conditioning unit adds on average about 50% to your annual electricity bill.

Air conditioning (the cooling of your office air to make it slightly colder than the outside temperature) is extremely energy hungry; it adds to your electricity bills as well as increasing greenhouse gas emissions.

Don't switch it on; reduce the need for air conditioning by:

Reducing internal heat sources Most of the energy consumed by conventional light bulbs is released as heat, which is very inefficient. Change your light bulbs to low-energy ones – remove this heat source and reduce your energy bill at the same time (see **Lighting** below).

Most conventional office equipment, from computers to photocopiers, also produces heat when operating, further warming your office air. Switch machines off when not needed or put them on stand-by. Consider using the 'traffic light system' for your office equipment: red – do not turn off; amber – put on stand-by; green – turn off (see **Office Equipment**).

Shading your windows Preventing the sun's rays from entering your office will help prevent the air in your office from warming up. Curtains and blinds inside the office help considerably. However, it is most effective to shade the windows externally by fitting shutters or awnings, which can be put away when not required.

If your office is on the ground floor, consider creating shade by growing plants around, above and over your windows. There are many fast-growing beautiful plants, which will add a new dimension to your office space and be a pleasure to watch throughout the seasons (see **Plants** below).

Ventilating naturally – open the windows! Experiment with opening various windows to produce a pleasant cooling breeze; on all but the stillest day it is possible to generate natural ventilation.

Air Fresheners – *see also Aerosols*

Open windows rather than using air fresheners, as many contain potentially dangerous chemicals that could damage your health. Locally grown flowers and plants provide natural fragrances and colour in the office – a much more pleasant alternative.

Air Travel – *see Flights*

Ball Point Pens – *see Pens*

Batteries

Used batteries are hazardous waste and must not go to landfill. Have a collection point for batteries and find out from your local council how to recycle or dispose of them.

Bleach – *see also Cleaning, Kitchens*

Avoid using chlorine-based bleach, including toilet blocks and household bleach, as it can cause damage to health and the environment.

Bicycles – *see Cycling*

Binders – *see also Office Equipment*

Most binders are only used infrequently and therefore should be switched off when not in use. Put a 'green' sticker on them if you adopt the traffic light system group (see **Office Equipment**). However if your office uses a thermal binder frequently and it takes a while to warm up, then it might warrant an 'amber' traffic light symbol.

Choose a comb binder which does not need electricity if possible.

Buses

Buses are one of the most energy-efficient ways to travel – use them where possible, especially between cities when you can get some really cheap deals.

Calendars

Why not use a computer programme for your planning. Most have calendar software programmed in; upload to a web page if others need to access it.

Calculators

Use solar-powered where possible, rather than mains-operated.

Cans

If aluminium drinks cans are used in your office, make sure you have a recycling scheme for them; find out if there is a 'cash for cans' centre in your area, or a charity that would welcome the money it could raise from your donated aluminium. **www.alupro.org.uk**

Carbon Dioxide

Carbon Dioxide (CO_2) is the main greenhouse gas contributing to climate change. Levels of CO_2 in the atmosphere are rising dramatically, and we urgently need to reduce our CO_2 emissions in order to avoid catastrophic climate change.

Carbon Footprint

Your carbon footprint is the measure of the amount of carbon dioxide your activities add to the atmosphere. Carbon Dioxide (CO_2) is the main greenhouse gas contributing to climate change. Surprisingly, many items – from apples to cars – can have a carbon footprint too, especially if they have been flown thousands of miles or if energy has been used in their production. Your purchasing choices affect your overall carbon footprint.

Carbon Offsetting

Carbon Offsetting is the principle whereby the carbon emissions created by activities such as flying or driving can be theoretically 'offset' by donating money to various 'green' projects such as tree-planting or renewable energy schemes. The notional CO_2 that these projects possibly 'save' from going into the atmosphere or reduce at some possible future date, gives the perception that people can carry on polluting and buy their way out of the problem. It is preferable to cut emissions in the first place, and only to consider credible carbon offsetting as a last resort.

Cardboard – *see Packaging*

Car Fleets – *see Fleet Cars*

Carpets – *see also Flooring*

If you are replacing a carpet, consider purchasing one made from natural materials that come from a renewable source rather than a man-made fibre derived from oil. Examples such as wool and sisal, jute and coconut coir not only look and smell good but are hard-wearing too. In non–carpeted areas, such as the kitchen, use eco-friendly flooring: cork tiles wear well and insulate the floor. Does the carpet really need replacing – could it be steam cleaned instead?

Cars – *see also Cycling, Company Cars*

Road transport contributes over 22% of our CO2 emissions within the UK, with half of that coming from our cars.

We urgently need to use our cars less, thereby reducing CO2 and other polluting emissions and reducing road congestion, as well as saving ourselves money and getting fitter in the process.

Many of us are wedded to our cars, and require considerable enticements to leave them at home and come to work another way. But once car owners are aware of the personal, social and financial advantages, many become passionately committed to greener ways of travelling.

More than half of all car trips are under 3 miles, which is about 10-15 minutes by bicycle

There are many advantages for both staff and the company to promoting greener ways to commute, such as:

- Saving money, and in some cases even earning extra
- Reducing the need for car parking space
- Reducing business miles
- Having healthier and more motivated people
- Reducing CO_2 emissions

Your objectives will be to reduce single-car occupancy and promote alternative forms of transport, both for commuting to work and for other business trips. This is normally achieved through workplace travel plans which, by means of plenty of carrots and maybe a few financial sticks, are very successful.

There are a number of organisations that exist to help companies to both formulate and implement a travel plan; there are travel plan conferences, workshops and websites to share best practices. The Association for Commuter Transport is a good place to start. **www.act-uk.com**

There are many things you can do to encourage people to switch to alternative means of commuting:

- Subsidise bus/rail/tube tickets. The costs will be significantly less than the annual costs of maintaining car parking spaces
- Provide easily accessible information about public transport with maps showing cycling and walking routes, timings, timetables and contact names via your green notice board, email, your Local Area Network (LAN) or intranet.

- Approach your local bus company (armed with facts and figures) to see if they will re-route a bus at key times if your office is not well served by public transport
- Create good facilities for cyclists, secure bike storage, showers with lockers, free breakfasts, and financial incentives for cycling to work
- Set up a 'bike buddy' and/or 'walking buddy' database through email, your Local Area Network (LAN), the company website/intranet, or notice board
- Organise bike purchasing through your company, thereby saving about half the total costs of the bike
- Set up weekly visits by a 'bike doctor' who will repair and maintain bikes

For those who have difficulty changing their means of transport, you can reduce the number of cars coming to the office by:

- Promoting 'car-share' schemes, with priority parking spaces for those who participate
- Setting up a 'car-share' database on the company website/ intranet, Local Area Network (LAN), by email, or on a notice board for potential car sharers with a guarantee of a free taxi should their lift not turn up
- Reviewing your business mileage allowance for cars. Set a flat rate for all cars based on the most fuel-efficient, regardless of engine size
- Organising occasional car-free days, with incentives and rewards

www.liftshare.org www.carshare.com

Reduce the need to travel:

- Assess potential conference venues and prioritise according to accessibility by public transport

- Review your business travel – is the visit really necessary?
- Investigate using video conferencing – very cheap with PC mounted cameras, and it can be done from any workstation
- Ensure that all visitors are made aware of public transport links and of your policy to reduce car use
- Can some employees work from home? Just once a week will substantially reduce staff journeys
- If people get in their cars to go and buy their lunch, consider a canteen, or find a sandwich delivery company, or organise a bus if you have sufficient numbers

Car Purchase – *see Company Cars*

Car Sharing – *see also Cars*

If you have to take your car to work, can you share your care with others making a similar journey?

Celebrations – *see also Christmas*

When you are ready to celebrate your success, make sure you do so in an environmentally friendly way, for example by using:

- Locally grown flowers
- Real glasses rather than plastic ones (often supplied free by drinks suppliers)
- Locally sourced food and drink

Chemicals – *see also Cleaning, Furniture, Glues*

Reduce the amount of toxic chemicals in the office where possible, whether in cleaning materials, furniture, fabrics, glues or inks. There are many environmentally friendly products available.

Christmas – *see also Celebrations*

Get a Christmas tree in a pot which can be planted in a garden and used the following year. Recycle all greetings cards, send email cards, or even make your own office cards. Buy 'green' presents, and make sure as much as possible of any food and drink you buy is organic or locally sourced.

Christmas Cards – *see Christmas*

Cleaning – *see also Aerosols, Chemicals*

Products used to clean offices can be toxic, with chemicals and solvents that could damage your health (for example bleach) and pollute the environment; check out the ingredients in your soaps and detergents, and avoid phosphates – there are many environmentally-friendly cleaning products available.

Where there is a cleaning contract, this can be reviewed for opportunities to change to products that are safe for the cleaners, those who work in the office, and the environment .

Coffee – *see also Organic, Tea*

Buy and use organic coffee if possible – less carbon dioxide is emitted in organic farming. Organic produce is readily available and promotes a much more pleasant and safe way of farming; and how about using organic milk with your coffee?

If you use 'real' coffee, why not compost the coffee grounds afterwards?

Commuting – *see also Bicycles, Buses, Cars, Car Sharing, Trains*

Leave your car at home; use public transport, walk or cycle to work instead. See **Cars** above for more information.

Compact Discs

Use re-writable CDs in preference to non-reusable CDs. Can you use a memory stick or external hard drive? Recycle your CDs and DVDs. **www.polymer-reprocessors.co.uk**

Company Cars – *see Cars, Fleet Cars*

Composting

Instead of throwing all left-over food, tea bags and coffee grounds into the waste bin, why not set up a composting system to prevent these organic materials going to landfill? When these materials decompose in a landfill site they produce methane, an extremely potent greenhouse gas. Even if your office is not on the ground floor, there are various methods now available to allow you to compost your organic waste. The keen gardeners in your office will eagerly take the finished product home.

Your local council or water company may sell you a composting system at a discount. If you want to be able to compost all types of food waste, including meat and fish products, make sure you purchase the correct type – tell your council or garden centre what you want to be able to put into it. There are many different composting systems for different purposes, and you will need to choose one according to you where you want to site it.

Some garden space required

- Tumblers – available through some councils and large garden centres **www.organicgardening.org.uk**
- Green Johannas – The 'Rolls Royce' of the plastic compost bin (the manufacturers claim you need no composting experience) **www.greenjohanna.se**
- Digesters such as the 'Green Cone' dispose of the waste but do not produce compost **www.greencone.com**

No garden space required

- Bokashi system – uses bacteria that thrive without air to ferment the material. No unpleasant smells are produced, and they can be used indoors. **www.livingsoil.co.uk**

- Wormeries – this system is great fun but requires a little more effort; the worms within it need looking after! However it is most rewarding. The worms eat food waste, paper and cardboard producing 'worm castes' a very valuable plant fertiliser. Once again, your local council might sell you one at a reduced price. There are many different types and sources. **www.greengardener.co.uk**

GETTING IN A PICKLE

Wiggly Wigglers take all their paper office waste, shred it and use it as a packing material, and use a Can-O-Worms outside the door for teabags, coffee grounds and other organic waste. Wanting to develop this, they decided to use a Bokashi Bin inside the office. As a result they can now compost banana skins, left over sandwiches, and flowers straight into the bin without any hassle (or indeed rotting vegetation). They set up two buckets so that one can be filled while the other pickles. There are absolutely no rotting smells, just a kind of pickle that they dig into the vegetable garden. For a team of nine people, they fill each bucket about once a month, and so emptying it is not a big deal – and their courgettes and radishes are seriously envied!

www.wigglywigglers.co.uk

Photo of Bokashi bin courtesy of Wiggly Wigglers

Computer Screens – *see Computers*

Computers – *see also Office Equipment*

A computer left on overnight uses 1Kwh of electricity; if 1,000 people turned off their computers when they went home, they would save 180 tonnes of CO_2 emissions every year.

Office computers tend to stay switched on, quietly consuming electricity throughout every working day; Surveys have shown that many office computers are never switched off, because either the operator can't be bothered or is unaware of the need.

- Turn computers off when not in use, especially at the end of the day. It's a great 'Easy Hit' for your green office campaign, and boosts morale and momentum. Just turning off computers at evenings and weekends will cut their running costs by more than two-thirds

- Set all computers to energy-saving mode so that the screen is switched off if the computer has not been used for more that a few minutes

- Don't be fooled by screen-savers – they use as much energy as the normal screen

- Set all computers to switch to 'stand-by' mode when not used for a short while. The power needed to restart is equivalent to only a couple of seconds normal running time

- When replacing computers, compare overall energy use (both running and stand-by). Check that they have stand-by or power-down modes

- Flat-screen or LCD monitors consume about one-third of the energy of traditional screens

    Computers are 'green' in the 'traffic light system' as they can be switched off when not in use – see **Office Equipment**

When upgrading computers and monitors, give old ones a longer life by selling or giving to staff for home use, or contact one of numerous commercial organisations or charities that will take your redundant computers, wipe the hard drive clean and either sell them or donate them to charities, schools or for use in developing countries. See **www.wasteonline.org.uk** to find a computer recycling organisation near you. **www.computer-aid.org**

Less than 20% of PCs in the UK are currently recycled.

Correctors

Buy correctors that do not contain toxic solvents; look for makes that are trichloroethane-free.

Couriers – *see also Deliveries, Post*

If you use or are considering using couriers to move both paperwork and light objects around a city, consider using a cycle courier: they are the ultimate environmentally-friendly couriers – they have zero carbon emissions, and are quick and efficient.

For larger parcels or for greater distances check out courier companies that use bio-fuels or are endeavouring to make your delivery carbon neutral by supporting alternative energy and tree-planting projects. Avoid courier companies that use air freight.

Credit Cards

Can your company use a credit card linked to an environmental charity?

Cups – *see also Kettles, Kitchens, Vending Machines, Water, Water Coolers*

Use 'real' cups, mugs and glasses that can be washed up, rather than disposable ones. When washing up, wait until there are sufficient numbers of mugs to warrant filling a washing-up bowl. Use an eco-friendly washing-up liquid, and try to use as little water as possible. Don't wash up under a running tap. Consider purchasing an 'AAA' rated energy efficient dishwasher if your office and kitchen is large enough.

If your office uses a water cooler with disposable cups, see if glasses can be used instead.

Curtains

Curtains are good heat insulators, especially if lined with thick materials such as brushed cotton. They can provide shading from direct sunlight during hot summer days and keep heat in during cold winter days and evenings. Draw curtains over windows at night, as they provide insulation and help keep the heat in the room.

When replacing curtains, use natural renewable fibres such as cotton or wool rather than man-made materials made from non-renewable sources.

Cycling – *see also Cars*

Bikes are a great way to get to work – they are cheap, carbon neutral, easy to park, often quicker than cars, and good exercise.

Cycle to work instead of driving your car. Try it for a day a week to start with, and choose days when the weather is fine. You'll save yourself money and become fitter in the process. Break the car habit, reduce your carbon emissions and add years to your life!

Diaries – *see also Calendars*

Can you keep your diary on your computer rather than using a hard copy?

Deliveries – *see also Couriers*

If you have sufficient space, change your ordering so that you order occasional large consignments of consumables rather than ordering small amounts when you need them. The fewer the number of deliveries, the smaller your carbon footprint.

Detergents – *see Cleaning*

DVDs – *see Compact Discs*

Electricity – *see Energy Purchasing, Lighting, Heating, Office Equipment*

EMAS – *see also ISO 14001*

The Eco-Management and Audit Scheme is a voluntary initiative designed to improve companies' environmental performance by recognising and rewarding organisations that comply.

It is an internationally accepted standard of Environmental Management. By participating in this scheme your company will reduce its environmental impact, maintain profitability and attract new customers. **www.emas.org.uk**

Energy – *see also Air Conditioning, Computers, Energy Purchasing, Heating, Kettles, Lighting, Office Equipment, Photocopiers, Printers, Scanners, Ventilation, Windows*

There are major financial and environmental benefits in reducing your energy consumption. Energy is used to heat your office in the winter, and possibly to keep it cool in the summer. Energy in the form of electricity runs all your office equipment, lifts, lighting and ventilation. It is a major office expenditure, and can easily be reduced by a few simple actions. See individual topics for specific information. **www.est.org.uk**

Energy Purchasing

Reduce the carbon emissions of your office at a stroke by changing your electricity provider to one which gets its electricity from renewable sources. For information on electricity companies offering green tariffs in your area, see **www.greenelectricity.org**

Energy Star

Where possible buy office equipment with the Energy Star label. This is found on office products that meet or exceed energy efficient guidelines. **www.energystar.gov**

Envelopes

Use envelopes made from recycled paper; these come in many colours and qualities, just like virgin paper envelopes and are available from most suppliers. If you use window envelopes, get those with the windows made from cellulose window film, as this can be recycled. Reuse envelopes whenever possible – open carefully, resealing and readdressing before use. Many charities sell envelope reuse labels, which can cover up the original address and stamp.

Environmental Champions – *see Environmental Taskforce*

Environmental Taskforce

Those members of the organisation who volunteer to lead the environmental improvements to the office (see page 19).

Envirowise – *see also* Global Action Plan

One of a number of organisations which are able to help you reduce your environmental impact and thereby increase your profitability. This government-funded organisation offers initial confidential visits, support packages and advice on grants and assistance. **www.envirowise.gov.uk**

Fairtrade

Primarily an ethical rather than green issue, Fairtrade products are purchased directly from the producers or growers, guaranteeing them a better price for their products. The extra premium on these products is reinvested by Fairtrade in social or economic development projects. **www.fairtrade.org.uk**

Fans – *see* Air Conditioning

Filing and Storage

Question whether you need to continue filing and storing paper copies of letters, invoices, reports, and suchlike. Instead of hard copies, back up electronic versions daily onto external hard drives or CDs. If you do need paper files or folders, buy recycled if possible.

Fleet Cars – *see also* Cars

- When leasing or purchasing fleet cars consider green issues such as CO_2 emissions and miles per gallon. A lower rate of emissions reduces car tax liability. **www.vcacarfueldata.org.uk**

- Keep your fleet maintained and regularly serviced

- Keep your fleet cars longer if possible – do they really need to be replaced so often?

- Give staff training to drive economically. Find out more at **www.rospa.com**

- Encourage staff who drive fleet cars to use public transport where possible and minimise their fuel consumption

Flights – *see also Foreign Travel*

On average, flying contributes about 10 times as much carbon dioxide to the atmosphere as a similar journey by train

Ensure that all staff are aware of the environmental damage caused by flying. Make it company policy that all longer business journeys are taken by train rather than plane where possible. Advance booking train journeys can give considerable savings. Or try to avoid making the journey altogether: investigate using video conferencing (prices for this facility have now tumbled with PC mounted cameras). **www.nationalrail.co.uk**

Flip Charts

Buy recycled flip charts. Can you reuse your used flip chart paper for office pads?

Flooring – *see also Carpets, Paints*

Natural products that come from sustainable sources such as wood, cork, bamboo and lino (not vinyl floor coverings, which are man-made) are preferable to man-made products, which pollute your office environment with unpleasant chemicals and are generally derived from oil. Ensure these natural floorings are coated with non-toxic products to keep your office healthy and environmentally friendly.

Flower Miles – *see also Gardens, Plants*

If you buy flowers for the office or for individual staff for special occasions, consider where the flowers have been grown and the method of transportation. The vast bulk of flowers bought in the UK are grown abroad and flown in from countries as distant as Columbia, Kenya and Israel. Flower air miles damage the environment, contributing to the carbon dioxide levels in our atmosphere and thereby climate change, and the way they are grown is very often harmful to those who work with them and the local environment. Consider purchasing flowers obtained from UK sources, and buy flowers that are in season rather than hothouse-forced ones. **www.wigglywigglers.co.uk**

Flowers – *see Plants*

Folders – *see Filing & Storage*

Food

Whether you are purchasing for a canteen or boardroom lunch, or simply want a snack, try and buy locally grown, organic produce where possible and support your local shops. Avoid highly packaged food.

Food Miles – *see also Food, Kitchens*

Food miles are the miles your food has travelled to get from its origin to your plate. The greater the distance, the greater the CO_2 created, and therefore the greater the impact of your food on climate change – especially if it has been freighted by air from its source. Local food, in season, both grown and sold locally, creates the minimum food miles.

However there are some occasions when certain types of 'out of season' food, such as tomatoes grown in Spain and freighted by road and ferry to the UK, create less CO_2 than those grown in heated greenhouses in the UK!

Buy locally-grown food in season and sold through local food shops.

Foreign Travel – *see also Flights*

Consider the train when travelling to meetings abroad. Air travel is particularly damaging to the environment, adding much more CO_2 to the atmosphere than any other form of travel. Now we have a high-speed train link to Europe, trains give a fast and environmentally-friendly way of travelling, are as quick as flying in many cases, drop you in the middle of a city, and your carbon footprint will be about ten times smaller if you travel by train rather than plane. When planning conferences, choose a destination that you can reach by train. Think trains when you plan your holidays too.

www.seat61.com (This site enables you to plan your travel worldwide using trains and ships).

Forest Stewardship Council (FSC) – *see also Furniture, Paper, Programme for the Endorsement of Forest Certification*

FSC is a widely used certification for timber and other products that originate from trees, which enables the purchaser to be confident that the forest source is managed sustainably and does not contribute to global forest destruction. **www.fsc-uk.org**

Franking Machines – *see also Stamps*

There are now alternatives to franking machines: you can print your own stamps from the web, which means one less machine running in your office which saves energy and money. **www.postoffice.co.uk**

Furniture

Furniture made from natural timber rather than chipboard or MDF (Medium Density Fibreboard) is preferable, as the latter leaches formaldehyde, a chemical which has the potential to damage your health.

When purchasing natural timber products, ensure that the timber used comes from a sustainable source. Look out for the FSC (Forest Stewardship Council) or PEFC (Programme for the Endorsement of Forest Certification) symbols. A good deal of office furniture is thrown out every year – consider buying second-hand or refurbishing your existing furniture, or donating it to the Furniture Recycling Network. **www.frn.org.uk**

Gardens – *see also Composting, Plants, Water Butts*

If your office is on the ground floor and you are fortunate to have some outside space, take full advantage of it. Grow wonderful plants, use them to shade your windows in the summer, and give your office space a third dimension. Pick flowers and foliage to put in your office – a bunch of freshly picked flowers brightens up everybody's day.

Global Action Plan – *see also Envirowise*

GAP is a practical environmental charity which can help you reduce your environmental impact and thereby increase your profitability. They offer support to companies wishing to green their offices and save money, and advice on grants and subsidies. **www.globalactionplan.org.uk**

Glues – *see also Solvents*

Buy and use vegetable or water-based glues; avoid glues containing toxic solvents such as toluene and xylene. Buy products such as these from companies that are well informed and only sell environmentally-friendly goods.

Ground Source Heat Pumps – *see Renewable Energy*

Heating – *see also Air Conditioning, Curtains, Insulation*

Simply by turning the temperature of your heating down by 1°C you can reduce your energy bill by 10%.

Heating your office during the winter months is a large proportion of your overall energy bill. The CO_2 generated in producing this heat contributes to climate change; however, there are many simple actions you can take in your office to reduce your energy use and therefore your carbon footprint.

- Consider turning down the thermostat controlling the temperature of your office by 1°C – you will probably not notice the difference. Research has shown that office staff are generally comfortable at 19°C, so turn down the thermostat incrementally to arrive at this temperature
- Turn radiators off or down in rooms that are only used occasionally, and turn the heating up when needed. Keep these rooms ventilated to prevent condensation and possible mould
- If your radiators are underneath windows, tuck any curtains in behind them to enable the heat to come into the room
- Move furniture away from radiators or heaters, to allow heat to get out into the office
- If possible make sure that your office heating system is set to come on about half an hour before the office will be used in the morning, and to go off about half an hour before the last person leaves in the evening
- Check all the timings, especially for weekends, when the heating can be on for considerably shorter periods and at a much lower temperature; 5°C is sufficient to prevent pipes bursting in cold weather
- If your office shuts down for a holiday break, adjust heating settings accordingly

- If your office is too hot in the winter months, turn the heating down or off rather than opening windows. If your heating is on, make sure all windows are closed

WARNING Don't block up air vents or grilles in walls if you have an open gas fire, a boiler with an open flue or a solid fuel fire or heater. These need sufficient ventilation to burn properly – otherwise highly poisonous carbon monoxide gas is released.

- Keep external doors shut, and if there are still draughts see whether draught excluders can be fitted round doors and over letterboxes; this won't cost very much. But make sure you still have sufficient ventilation – see above
- If there are draughts coming from under skirting boards or through floorboards, see whether gaps can be filled. But make sure you still have sufficient ventilation – see above
- Service your heating system regularly – it will be more efficient and use less energy, saving the company money and reducing your offices' contribution to climate change
- If your boiler is due for renewal, consider the most energy-efficient model rather than the cheapest available. As energy prices continue to rise, it will save money and produce less CO_2
- Make sure all radiators have individual thermostats where possible – this will allow you to vary the temperature in different rooms, according to use
- Use plug-in electric heaters such as bar heaters, fan heaters, oil filled radiators or panel heaters sparingly – they are very expensive to run

Home

Encourage staff to put into practice at home the environmental actions they are doing in the office. There are huge financial savings to be made in the home by taking some simple actions:

the less energy you use, the smaller your energy bills and the less CO_2 is released, benefitting us all by helping to reduce climate change.

Hydro Power – *see Renewable Energy*

Information Packs – *see Training Manuals*

Ink Cartridges – *see also Deliveries, Ordering*

It is relatively easy to forecast your usage of ink cartridges. Benefit financially and help save the planet by ordering occasional bulk orders rather than frequent small orders when needed. Consider using remanufactured cartridges, or getting them refilled. **www.sort-it.org.uk www.recyclethis.co.uk**

There are over 30 million inkjet cartridges thrown away every year in the UK. Most of these could be reused many times by being refilled or remanufactured.

Insulation – *see also Air Conditioning, Curtains, Heating, Windows*

If you are building a new office or refurbishing an existing one, ensure that all walls and the loft or roof spaces are well insulated. This will both prevent heat loss in the winter and keep you cooler in the summer, thereby dramatically reducing your heating and air conditioning costs.

Almost 40% of all the heat used to heat your office escapes through the walls and roof spaces if they are not insulated.

recycle
your empty
inkjet cartridges

WWF is best known for its work to conserve endangered species. But WWF also addresses other global threats to the planet, such as climate change and deforestation. Help WH Smith raise a target of £100,000 for WWF. At least 50p will be donated for every cartridge that is recycled.

You can use this bag for more than one cartridge!

Please use this bag for all empty Hewlett Packard, Lexmark, WH Smith own brand and Canon BC01/02/05/06/20/BX2/BX3
Please DO NOT SEND any other inkjet or laser cartridges.

To order
08

WWF Re,
FREEPOST L
LONDON
NW10 7ZZ

WWF. Taking action for a living planet

Insurance

There are a limited number of insurance companies that put a percentage of their premium into environmental projects and offer environmental performance reviews. **www.naturesave.co.uk**

ISO 14001 – *see also EMAS*

International Standards Organisation is an accreditation organisation for the UK. This is an internationally understood and accepted standard of Environmental Management. It will help your company to reduce its environmental impact whilst maintaining profitability, and enable you to attract new customers. **www.netregs.gov.uk**

Invoices

Consider using electronic invoices instead of paper: you can create your invoices as PDF documents and email them to customers.

Junk Mail – *see also Mailing Preference Service*

Send it back in the prepaid envelopes and asked to be removed from the database – trees have been cut down to create this waste. For your own marketing, why not use email?

Kettles – *see also Cups, Coffee, Fairtrade, Kitchens, Water*

Electric kettles are in frequent use, and consume a surprisingly large amount of electricity, which varies considerably according to the model. When you replace your kettle, choose one with minimum energy consumption. Your kettle will be more efficient if it is kept free of limescale.

Kettles often heat up more water than is necessary; only boil as much as you need. Many now have an indicator showing the amount of water they contain. Consider an insulated kettle, which will keep any extra water hot until next time.

If your office is large enough, investigate using a local water heater or urn instead of a kettle; you could save considerably on energy used and have hot water for tea and coffee instantly available.

Kitchens – *see also Cleaning, Composting, Cups, Coffee, Fairtrade, Kettles, Milk, Water*

- Set up well-labelled recycling bins in the kitchen for plastic bottles and cans, with another bin for all the packed lunch waste, tea bags and coffee grounds, which can go to be composted. To ensure success, pin up a poster to explain where things should go and why
- If the hot tap water is too hot to use, see whether the temperature can be turned down. Don't waste energy heating water only to have to add cold water so that you can use it! Just over 60°C should do it
- Make sure there are no dripping taps. One dripping tap can waste at least 5,500 litres of water a year
- For heating or defrosting food, a microwave is more energy-efficient than a conventional oven
- If your office is large enough, consider purchasing an 'AAA'-rated energy-efficient dishwasher rather than washing dishes and cups by hand. Providing the dishwasher is full and its economy or 'eco' mode is used, 'AAA'-rated dishwashers are surprisingly frugal with both water and energy usage
- Use environmentally-friendly cleaning products for both washing up and cleaning the kitchen area. There are several well-known brands available
- If you need kitchen rolls, buy recycled

Labels

Why not print directly onto the envelopes and minimise the use of labels?

Landfill

Space for landfill is fast running out. Recycle as much as you can.

Laminators – *see also Office Equipment*

Most laminators are only used infrequently, and consequently should be switched off when not in use. They will therefore have a 'green' sticker on them if you adopt the traffic light system group (see **Office Equipment** above).

However if your office uses a laminator frequently and it takes a while to warm up, it might warrant an 'amber' traffic light symbol. If replacing a laminator, buy one which shuts off after a period of inactivity.

Laptops

If you are replacing a computer, consider a laptop as they are much more energy-efficient.

Lifts

Don't waste electricity; use the stairs instead. Reduce your carbon footprint and get fitter in the process.

Lift sharing – *see also Cars*

Can you travel with someone else, share costs and save a journey? **www.liftshare.com**

Lighting

This is one of those 'easy hit' targets – with a simple 'Switch Off' campaign you can get most of the staff involved and motivated.

- Replace all conventional light bulbs with energy-efficient light bulbs, which last about 12 times longer than ordinary bulbs and consume about 1/5 of the energy.

- Label all light switches to say which lights they operate, with a reminder to switch off when not needed

- Encourage staff to use natural lighting when available

- Evaluate the office layout, and consider moving furniture to maximise natural lighting

- Don't use more lights than you actually need

- In areas not in constant use, such as toilets, hallways, stairs and storerooms, replace the light switches with movement sensors that switch on the lights only when needed. Timer switches can also be used where appropriate

- Make it clear that it is the responsibility of the last person who leaves the office to ensure all the lights are switched off when they leave. If this is the office cleaners, then raise this with them and their management

- Uplighters can use high wattage bulbs that are expensive to run – use energy-efficient spots instead

- Halogen bulbs are not energy-efficient; energy-efficient replacement bulbs are now available

- Contrary to popular belief, you save energy if you turn strip lights off when not in use – switch them off!

Energy-efficient light bulbs are cheap to run because they mainly make light rather than heat. 90% of the energy used by traditional bulbs is wasted in producing heat.

Are you the last person
to leave?

Don't forget

TURN OFF
ALL THE
LIGHTS

Local Area Network (LAN) – *see also Paper*

A LAN (office computer network), which enables you to share and send information electronically, is a key tool in the paperless office.

Mailing Preference Service

Register with the Mailing Preference Service to reduce your junk mail. **www.mpsonline.org.uk**

Marker Pens – *see Pens*

Milk – *see also Coffee, Kitchens, Tea, Fairtrade*

Use local organic milk – it's much better for both you and the planet, it will have travelled less distance than ordinary milk and you'll be encouraging a more sustainable and healthy way of farming.

Minicabs – *see Taxis*

Mobile Phones

If your mobile stops working, don't throw it away – get it repaired. There are many companies who will mend your mobile at a very reasonable price; simply post them your phone.

If it is beyond repair, recycle your mobile. Raise money for your favourite charity and divert your piece of electronic waste away from landfill. **www.mobilephones4charity.com**

In the UK, over one and a half thousand mobile phones are purchased every hour, and over eleven million are thrown away every year.

Monitors – *see Computers*

Mouse Mats

Buy mouse mats made from recycled materials. They are fun, cool, and give you the opportunity to flaunt your 'green' credentials.

New Building – *see also Air Conditioning, Carpets, Curtains, Flooring, Furniture, Heating, Insulation, Lighting, Paints*

If you are lucky enough to be considering building a new office, now is your chance to have an energy-efficient office tailor-made.

With a new build you have the opportunity to invest in renewable energy systems, superb insulation, well designed windows that allow sufficient natural light in but don't overheat the office in the summer, energy-efficient heating, natural cooling and ventilation, and so on.

Your Environmental Taskforce can play a key part in collating suggestions and ideas from other members of staff and presenting them to senior management. Although the initial build costs might be a little higher, these expenses will soon be recouped through much lower energy bills. This is also an ideal opportunity to benefit from some very positive PR.

Notepads

Make office notepads from paper already printed on one side. Cut them down to size and hold together with a bulldog clip.

Office Equipment – *see also Binders, Computers, Laminators, Laptops, Photocopiers, Printers, Scanners*

It is currently estimated that currently over 1 million tonnes of waste electrical equipment is discarded in the UK every year, and this figure is on the increase.

When purchasing electrical equipment, compare running costs and energy use whilst on stand-by. Check that it has stand-by or power-down modes and other energy saving features.

If your equipment is not too old, it can probably be refurbished and reused. Check out the internet for reuse and recycling centres near you.

TRAFFIC LIGHT SYSTEM

This system works well for electrical office equipment. Simple coloured stickers in a prominent position on all electrical machines show employees whether machines should be left on or turned off. Make people aware of the system by poster or email.

GREEN – use for machines that can be switched off when not in use – rather than letting them go into standby mode. Most office computers will be green.

AMBER – use for machines that are best left switched on during the day, as they take a long time to warm up, but can be switched off at the end of the day. Photocopiers are a good example of equipment that will be marked amber.

RED – is for equipment that has to be kept turned on all the time. Office answerphones and wax printers are typical red coded machines.

Order Confirmation

Use email rather than paper where possible.

Order Forms – *see also Invoices, Statements*

Consider using electronic order forms instead of paper; you can create them as PDF documents and email them to customers. You could also suggest to your customers that they use the same system.

Orders – *see Couriers, Deliveries, Order Forms*

Organic – *see also Coffee, Fairtrade, Kitchens, Tea*

Support organic farmers and encourage a more pleasant and healthy way of farming. Pesticides and fungicides sprayed onto produce are potentially harmful to the farmer, the planet and you the consumer. Reduce your exposure to such chemicals by choosing to eat and drink organic produce. Organic tea, coffee and milk are now readily available. **www.soilassociation.org.uk**

Packaging – *see also Paper*

- Organise your office with bins to collect all the packaging you receive, so that you can reuse when required. Collapse and flat-pack cardboard boxes until required for reuse

- If you receive more packaging than you can reuse, don't throw it away – donate your surplus to other local companies

- Use starch-based, biodegradable packaging (white foam-like extrusions) for your packing. They dissolve in water, are non-toxic and can be composted

- Avoid using expanded polystyrene and plastic wrapping – there is biodegradable cellulose wrapping available for shrink-wrapping if needed

- Reuse jiffy bags where possible, and if you purchase jiffy bags choose ones that are lined with recycled paper rather than bubble wrap, as these can be recycled easily

Paints

Encourage the use of natural paints and wood finishes for your office. These are completely non-toxic, and contain no solvents or other poisonous byproducts to pollute your office environment.

There is a vast array of environmentally-friendly natural paints and associated products available to choose from. **www.ecosorganicpaints.com www.ieko.co.uk**

Paper – *see also Forest Stewardship Council, Notepads, Packaging, Printing, Programme for the Endorsement of Forest Certification, Stationery*

Our consumption of paper in the UK continues to rise by about 20% every year.

Incredibly, and in spite of much publicity, we are still consuming ever-growing quantities of paper, so the paperless office has yet to take off. Both the cutting down of trees for the production of virgin paper (which involves the use of a lot of energy and chemicals) and the disposal of paper waste, by either incineration or landfill, have high environmental costs.

The UK is one of the highest consumers of paper in the world, using more than 12 million tonnes of paper and cardboard every year, with this massive usage continuing to rise by about 20% annually. About half of all waste from an average office is paper. It is also a major purchasing cost, with many office workers using up to 100 sheets of paper daily. Reducing the paper consumption of your office will therefore have large environmental and financial benefits. It is also another 'easy hit' target which, with an email and poster campaign, can get most of the staff involved and motivated, and will produce quick, tangible results.

Reduce consumption

- Use email, notice boards, LAN or intranet rather than paper to communicate. Remind everybody not to print out emails unless absolutely necessary

- Encourage staff to use both sides of every sheet of paper. This should be possible for all printing and photocopying, apart from letters to clients

- Place posters or signs beside all printers and photocopiers reminding everybody to use both sides of the paper together with a 'Do you really need to print this?' sign

- Put a recycle bin beside each machine to collect paper that can be used again; encourage this to be used in preference to virgin paper

- Edit standard letters to fit onto one page if possible

- Check out whether you could install a fax server. This would allow faxes to be sent directly from your PC, thereby avoiding the need to print out a copy. You can even set up your fax so that incoming faxes come into your PC digitally. You can then decide whether you need to print them out or simply forward the fax electronically to those concerned

- Make sure your fax is set up so that it does not produce unwanted report sheets

- Check out 'Greenprint', a software programme that analyses what you intend printing, highlights and removes unwanted pages and lets you decide what you really want to print. It also enables you to make PDFs at the click of a mouse thereby possibly removing the need to print **www.printgreener.com**

- Print your own letterheads and reduce the need to carry stock of letter-headed paper. Company changes can be easily accommodated this way

- Use lighter weight paper where possible – less energy and material are used in its manufacture

Recycle waste paper and reduce your disposal costs

- If you don't have your waste paper collected for recycling, contact your local council offices or ring Envirowise on 0800 585794 to find which companies collect paper for recycling near you. Costs and methods of collection vary, so clarify these before signing up

- Paper recycling bins need to be readily accessible to all staff. One beside every printer and photocopier is a good idea, and one additional bin for every six staff can be distributed around the offices

- As with all schemes, its success depends upon ensuring that everyone is aware of what is happening and knows what to do. Email circulars and posters beside the recycling bins providing explanations will help avoid problems

- Staff feedback is essential: use all means available to provide the opportunity for people to make suggestions and to report on progress and problems

- Ensure that all staff are kept informed of recycling progress; graphs and figures are a great way of saying 'Look how well we're doing'. For more suggestions, see page 19

To make a tonne of virgin paper you need as much electricity as an average household uses in a year.

Purchasing Paper

- Try to buy 100% recycled paper wherever possible, depending on your print requirements.

- If you need to reassure staff that recycled paper compares favourably with virgin paper, get some samples of different types of recycled paper and try them out before bulk buying

- Try to purchase recycled paper that has the highest 'post-consumer waste' content. Terms you will come upon when ordering recycled paper include:

- 'Post-consumer waste' – waste paper that has been collected from offices and homes to be recycled
- 'Pre-consumer waste – scrap paper created by paper mills and other paper processing operations

- Endeavour to use recycled paper that has been produced in an environmentally sensitive way. In particular, try to avoid the use of chlorine (used to bleach the pulp) as is particularly damaging to the environment

- If you cannot get recycled paper to the specification you require, consider using FSC paper (see **Forest SC** above)

- To discuss and order recycled paper, see **www.paperback.coop**

The Paperless Office

You can reduce your paper use by:

- Using electronic methods for standard office procedures e.g. accounts, invoices, statements, order forms – PDFs are generally used

- Filing documents electronically, backed up as necessary

- Encouraging others to reduce their paper use by putting a message at the end of your emails asking the recipient to avoid printing it if possible

- Using electronic diaries and calendars

- Using your website to post documents for the general public and the media, e.g. reports, news

- Using a Local Area Network (LAN) to communicate with other members of staff

- Sharing information securely over the internet, for example using Googlemail documents and spreadsheets

Paper Towels – *see Kitchens, Toilets*

Paperclips – *see Staples*

Pencils

There are pencils made from recycled materials, or get a pencil for life – i.e. one which has replaceable leads

Pens

Why not give a fountain pen a try? Those which don't use cartridges are the most eco-friendly, and can last for many years. However, for those who still prefer the instant, non-smudge 'biro'-type pens, there are now huge ranges of pens and pencils, including marker pens, that are made from recycled materials and are biodegradable.

Ensure that the ink in the marker pens is non-toxic; it should be toluene-free and xylene-free. Check out refillable versions, and whether the supplier collects the 'used and empty' pens for recycling.

Photocopiers – *see also Office Equipment, Printers, Traffic Light System*

Most photocopiers can be included as part of the 'amber' traffic light system group (see **Office Equipment**) as they are in frequent use and normally require a period to warm up. It's therefore generally best to leave them switched on during the day and switch them off before leaving for home. Photocopiers, like printers, are often used indiscriminately, and they waste paper. Put a bin for scrap – i.e. paper which has been printed on both sides – and a box for paper which can be reused by the side of all photocopiers, and a couple of simple notices will help reduce paper use – 'Do you really need a paper copy?' or 'Use both sides of the paper if possible'.

Planners – *see Calendars*

Plants – *see also Composting, Flowers, Gardens, Water Butts*

Use plants in your office to 'green' your workplace. Plants have been found to be very effective at reducing air pollution in enclosed places (e.g. the spider plant (*Chlorophytum comosum*); they also act as excellent room dividers, as well as raising the humidity of the air and deadening noise. Pot plants last longer than cut flowers; buy plants that have been grown locally rather than flown in from abroad. Plant air-miles damage the environment, contributing to the carbon dioxide levels in our atmosphere and hence to climate change.

If you have some outdoor garden space, you can use this to grow your own office plants. Many benefit from a colder period outdoors prior to being brought inside, where they will flower, bringing colour and 'a breath of fresh air' to your office.

To avoid death by over-watering, make sure that one person is responsible for the care and maintenance of the office plants.

In 2006 19,000 tonnes of flowers were imported from Kenya to the UK, creating 33,000 tonnes of CO_2 emissions.

Plastic – *see Celebrations, Christmas, Cups, Packaging, Pens*

Waste plastic is a major problem for the environment, and much of it ends up in landfill. It is derived from oil, a non-renewable source. Offices traditionally use a lot of plastic product – for example shrink-wrapping and polystyrene packaging, both of which can easily be substituted with an environmentally-friendly product. Where possible, use biodegradable products made from a renewable source, or those that can be reused, rather than plastic.

Post – *see also Couriers, Deliveries, Stamps*

Whilst it is outside your control how the post is distributed within the UK, when posting items abroad consider whether they need

to go by air mail, or if they could be send surface mail. Although this takes longer, surface mail (ships or trains rather than planes) will reduce the carbon footprint of your posted items.

Posters – *see also Environmental Taskforce*

Judiciously placed posters around the office are an excellent means of informing staff about environmental schemes, including successes, stories, media coverage, future events and suchlike. If used in tandem with email circulars, your Local Area Network (LAN) or the intranet, they provide frequent reminders and 'keep people on board'. However, ensure that the environmental taskforce member responsible for posters keeps them alive and current. Tired 'dog-eared' posters can easily give the wrong message out to the staff. Fresh, lively posters with personal stories that seek to change habits by using a carrot rather than a stick are preferable. But beware of their overuse – this could be counterproductive, as the posters themselves require paper and printing.

Printers – *see also Binders, Computers, Ink Cartridges, Laminators, Office Equipment, Paper, Photocopiers, Scanners*

Most printers can be included as part of the 'amber' traffic light system group (see **Office Equipment**) as they are in frequent use and normally require a period to warm up. It's therefore generally best to leave them switched on during the day and switch them off before leaving for home. However, some wax printers are 'red' coded machines and need to be kept switched on all the time, so check this out before placing the appropriate 'traffic light symbol' onto the printer.

Encourage staff to utilise the 'economy' or 'speed' facility available when instructing the printer to operate from PCs (generally found under 'print quality'). These modes use considerably less ink and are much quicker. Save 'best' print quality for external communications.

See **Paper** for more information on how to reduce paper usage when printing.

Purchasing

When upgrading your printer, check to see that stand-by or power-down modes and other energy-saving features are included. Compare running and stand-by energy use before purchasing. Also check out the purchase price of the ink cartridges; see if they can be refilled or if remanufactured replacements are available.

Reuse and Recycle

Check out the internet for reuse and recycling organisations near you. If your printer is not too old, it can probably be refurbished and reused elsewhere.

Programme for the Endorsement of Forest Certification Schemes (PEFC) *see also Forest Stewardship Council*

This is a global umbrella organisation for certification schemes for timber and other tree products. **www.pefc.org**

Promotional Literature *– see also Training Manuals*

Where possible use email and the web to promote your business to save paper.

Public Transport *– see Cars*

Purchase Orders *– see Order Forms*

Recycling *– see individual subjects*

Recycling Bins – *see also Cups, Envelopes, Kitchens, Packaging, Paper, Rubbish*

Recycling is easy, but also needs to be made easy: place bins, clearly labelled, where they are needed and easily seen. For example:

- The kitchen and dining areas: check with your local council or recycling collection company to see what they collect and how they would like it separated. A bin for metal drinks and food cans, with another to collect certain types of plastics, would be great here
- Entrance lobby: the same types of bins as in the kitchen area can be placed here ready for staff back coming from lunch
- Next to printers and photocopiers: waste paper bins, again clearly marked as such but with reminder signs such as 'Have you used both sides?' placed on the bin. (See **Paper** above for further information)
- In shared office space: one waste paper bin for about every six staff, again with similar reminders as above

Renewable Energy – *see also New Buildings*

Renewable energy is energy produced by a source that continually renews itself; main sources are the sun, moving water, wind and plant materials. This energy can be used for space heating and hot water heating, and to produce electricity for your office. By using renewable energy instead of conventional energy sources you will reduce your carbon footprint and thereby lessen the impact of climate change.

Solar power

Energy from the sun can be used both to provide domestic hot water and to produce electricity for your office. Different technologies are used for each.

Wind turbines

Wind turbines convert moving wind into electricity. For many offices in the UK a new kind of micro-turbine that attaches to your chimney or roof is the most convenient and practical.

Biomass (biofuels)

Biomass or biofuels are materials such as wood or straw which grow quickly and can be burnt to release heat for space heating and domestic hot water. Biomass is different from all other renewable energy sources because the fuel generally has to be purchased.

Biomass is a renewable energy source because the materials are quick to grow, absorbing CO_2 in the process; and the CO_2 released when it is burnt balances that which was absorbed during the growth of the material, effectively making the process carbon-neutral.

Ground source heat pumps

Heat pumps take heat from under the ground (which remains at about 12°C all year round) and use it to heat your office – just like a refrigerator in reverse. You will need sufficient space outside to dig either a trench or a borehole if you want to install a ground source heat pump.

Small-scale hydro power

If you are fortunate enough to have a fast moving stream or river running near your office, it may be possible to generate electricity from the moving water. Such hydro schemes have the capacity to generate substantial amounts of electricity, which can then be sold back to your electricity company.

Reports

Send reports by email and/or put them on your website and invite people to download them.

Rubbish – *see also Recycling Bins*

An environmentally-friendly office will have minimum waste to dispose of in landfill. Much can be recycled or reused (see the appropriate subject). Sensible purchasing will dramatically reduce waste and therefore disposal costs. There are opportunities to discuss ideas and issues with other similar businesses at your local Resource Efficiency or Waste Minimisation Club. Contact Envirowise to see what's available in your area (see page 13 above).

- Ensure that the volume and/or weight of waste for disposal is noted and communicated to staff. Although there are some waste products such as sanitary waste that will not diminish, it should be possible to dramatically reduce all other products requiring disposal
- Keep staff on board by informing them of goals attained, future targets and possible prizes for high achievers
- Although it may seem slightly Orwellian, occasional bin checks when the office is empty will inform the Environmental Taskforce of weaknesses in the system

Scanners – *see also Office Equipment*

Most scanners are only used infrequently and therefore should be switched off when not in use, and should have a 'green' sticker on them if you adopt the traffic light system group (see **Office Equipment** above).

However, if your office uses the scanner frequently and it takes a while to warm up, then it might warrant an 'amber' traffic light symbol.

Shredders

Ideally, use a manual shredder; you can use the shredded material that is not confidential as packaging.

Soap – *see Cleaning*

Solar Energy – *see Renewable Energy*

Solvents – *see also Cleaning, Furniture, Glues, Pens*

Toxic solvents which evaporate into the air are present in a surprising number of office products, and can damage your health. There are now environmentally-friendly alternatives available – see individual topics for further information.

Stamps

Consider buying your postage online instead of using 'stick on' stamps. **www.printstamps.co.uk**

Staples

If all the office staff in the UK used one less staple a day, it would save 328 kilos of steel every day.

Staples are such small pieces of steel that they are easily ignored. Have a look at staple-less paper joiners, or use reusable paper clips instead.

Statements

Consider using electronic statements instead of paper; you can create your statements as PDF documents and email them to customers.

'Stick on' Notes

Use electronic versions, recycled paper notepads, or buy recycled.

String

Use biodegradable (e.g. cotton, sisal or jute) rather than plastic where possible.

Switch Off – *see also Lighting, Office Equipment*

This is the mantra of energy saving: switch off lights, heating, air conditioning and other electrical products when they are not in use. See the individual products listed for further information, and check out the 'traffic light system' – see **Office Equipment**.

Tax Returns

Do them on-line, saving paper and money.

Taxis

Like all other cars, taxis emit CO_2 – can you use public transport instead?

Tea – *see also Coffee, Composting, Kitchens, Milk, Organic*

Less carbon dioxide is emitted in organic farming – buy organic tea and use organic milk, ideally from a local or UK source.

If a number of you make tea at the same time, how about using a teapot and brewing up using loose tea rather than tea bags. This way you will avoid the additional packaging required for tea bags and possible effects on your health from the bleaching agents used by some manufacturers to whiten the bags. If you still need to buy tea bags, then purchase the non-bleached variety.

Don't forget: when you've finished your tea you can compost the used tea leaves and tea bags. See **Composting** above for further information.

Telephone Directories – *see also Yellow Pages*

Use the web to find telephone numbers. Be sure to recycle your old directories. Opt out of having telephone directories delivered by phoning 0800 833 400.

Telephones

Try to avoid base stations, which have to be charged. There is currently a debate about the health issues surrounding cordless phones.

Tissues

Bring back the handkerchief!

Toilets – *see also Air Fresheners, Cleaning, Water*

Toilets and hand basins make up the majority of the water used in an office, and your office water bill will be a major expenditure. Simple water efficiency measures allied with effective communications can reduce the water bill of your company by 50%.

- Install waterless urinals
- Install dual flush toilets. The short flush will be used most of the time
- You can reduce the amount of water that your existing toilets use by fitting water-saving devices such as 'Hippo', 'Save-a-flush' or 'Hog Bag' in the cisterns. These are normally provided free by water companies
- Install push-button taps on the sinks. These will cut your sink water usage by over 50%
- If your office has urinals which are not waterless, reduce your water consumption by fitting automatic flush controllers. The cisterns will then only flush during office hours or after use
- If you use paper towels, buy recycled
- If you use electric hand dryers, position them so that they are not turned on by passing people
- Avoid air fresheners; use natural ventilation instead

Discuss with the company that cleans your office the use of environmentally-friendly cleaning products for both your toilet areas and the remainder of the office. Make sure that water colouring/foaming agents are not placed in the cisterns – they are totally unnecessary and require additional treatment at the sewerage works.

Toner Cartridges – *see Ink Cartridges*

Traffic Light System – *see Office Equipment*

A system which can be used for all electrical office equipment:

> RED = must be left turned on
> AMBER = put on stand-by
> GREEN = switch off when not in use

Training Manuals

If your company uses training manuals, put them on the LAN or intranet as a download or make them available on a CD. Phase out the use of hard copies, as they consume large quantities of paper and are soon out of date and then need reprinting. If you have to print hard copies, try to get them printed on recycled paper using vegetable inks and avoid having plastic laminates on the covers.

Trains – *see Foreign Travel*

Use them!

Transport – *see Cars, Company Cars, Cycling, Flights, Foreign Travel*

Travel Plans – *see also Cars, Bicycles*

A travel plan is a scheme designed to reduce single car occupancy and promote greener forms of transport, both for commuting to work and for other business trips. **www.act-uk.com**

Vending Machines – *see also Cups, Water Coolers*

Vending machines are left on all the time, quietly using energy; food and drink in these machines is almost never from a local source and therefore involves many food miles. Look into alternative ways of providing food and drink for the office, such as a sandwich delivery service or a kitchen area. If you do lease a drinks vending machine or water cooler, try to get one which takes 'proper' cups or glasses rather than disposables.

Ventilation – *see Air Conditioning, Windows*

Waste – *see Rubbish*

Water – *see also Kitchens, Toilets*

Water is a significant proportion of your office expenditure, so reduce your water use and save money. Treating and pumping water to your office also requires huge amounts of electricity, so you reduce emissions of greenhouse gases responsible for climate change by using less water.

The majority of water used is in the toilets and hand basins, as well as in your office kitchen. See **Toilets** and **Kitchens** above for detailed information on how to reduce water consumption in these areas.

A few simple water efficiency measures and effective communications can cut your water bill by 50%.

Water Butts

Water down the drain is wasted; if your office has a garden or outside space, collect free water from the roof by installing a water butt, and use this water for your plants inside and outside rather than using treated drinking water. Some water companies and councils sell water butts at discounted prices.

Water Coolers

These are similar in design to fridges and use considerable amounts of electricity, as they are never turned off. There are two types – those which use delivered bottles of water and those which are connected to the mains water; avoid the additional CO_2 emissions associated with water miles and use the mains variety. When renewing or starting a lease for a new water cooler, check out the electricity consumption of the various models on offer and choose accordingly.

Wi-Fi

Although wireless networks are growing in popularity due the lack of cabling and ease of movement of equipment, there are concerns about long-term health implications of exposure to low level electromagnetic radiation from wireless networks. Consider using conventional cabling for your office networks.

Wind Turbines – *see Renewable Energy*

Windows – *see Air Conditioning, Heating, Ventilation*

Xmas – *see Christmas, Christmas Cards*

Yellow Pages – *see also Telephone Directories*

Use the web to find telephone numbers. Be sure to recycle your old directories. Opt out of having Yellow Pages delivered by phoning 0800 671 444. **www.yellgroup.com**

Resources

Association for Commuter Transport (ACT) offers advice, training and help to organisations considering a Travel Plan to encourage their staff to come to work by a sustainable means of transport. www.act-uk.com

The Aluminium Packaging Recycling Organisation (Alupro) is a not-for-profit company which provides bins and collection facilities for aluminium cans. www.alupro.org.uk

The Carbon Trust helps business and the public sector cut carbon emissions, and supports the development of low carbon technologies. www.carbontrust.co.uk

The Eco-Management and Audit Scheme (EMAS) is a voluntary initiative designed to improve companies' environmental performance by recognising and rewarding organisations that comply. www.emas.org.uk

The Energy Saving Trust (EST) provides independent advice and information about energy efficiency, insulation and renewable energy options and availability of grants. www.est.org.uk

Energy Star is a scheme that accredits office products that meet or exceed energy efficient guidelines. www.energystar.gov

Envirowise offers UK businesses free, independent, confidential advice and support on practical ways to increase profits, minimise waste and reduce environmental impact. www.envirowise.gov.uk

Fairtrade is an independent consumer label which guarantees that disadvantaged producers in developing countries get a better deal. www.fairtrade.org.uk

The Forest Stewardship Council (FSC) promotes responsible management of the world's forests, its gives accreditation to products manufactured from virgin materials sourced from well managed forests. www.fsc-uk.org

THe Furniture Re-use Network (FRN) is the national co-ordinating body for 400 furniture and appliance re-use and recycling organisations in the UK that collect a wide range of items to pass onto people in need. www.frn.org.uk

Global Action Plan (GAP) is an environmental charity which can help you reduce your environmental impact and thereby increase your profitability. They offer support to companies wishing to green their offices and save money, and give advice on grants and subsidies. www.globalactionplan.org.uk

Liftshare enables both drivers and passengers to find travel companions and share travel costs. www.liftshare.org

The Mailing Preference Service helps individuals and organisations to reduce junk mail. www.mpsonline.org.uk

The Man in Seat Sixty-One helps you to plan your worldwide travel using trains and ships instead of flying. www.seat61.com

The Recycle Now campaign is an independent, government funded organisation which provides advice to both individuals and companies on recycling. www.recyclenow.com

The Soil Association is the UK's leading campaigning and certification body for organic food and farming. www.soilassociation.org.uk

The Vehicle Certification Agency (VCA) helps anyone buying a new car to choose a model which has the least impact on the environment and provides information on new car fuel consumption and exhaust emissions. **www.vcacarfueldata.org.uk**

WasteOnline provides information on sustainability, resource use and waste issues. **www.wasteonline.org.uk**

Also in the Green Books Guides series:

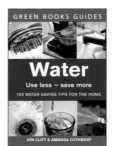

Water: use less – save more
by Jon Clift and Amanda Cuthbert
100 water-saving tips for the home, in full colour.
£3.95 paperback

Energy: use less – save more
by Jon Clift and Amanda Cuthbert
100 energy-saving tips for the home, in full colour.
£4.95 paperback

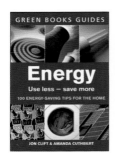

Reduce, Reuse, Recycle:
an easy household guide
by Nicky Scott
An easy-to-use A–Z household guide to recycling.
£4.95 paperback

Also in the Green Books Guides series:

Composting:
an easy household guide
by Nicky Scott
Tells you everything you need to know for
successful home composting.
£4.95 paperback

Cutting Your Car Use
by Anna Semlyen
Tackle car dependency and change your
travel habits.
£4.95 paperback

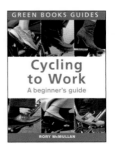

Cycling to Work: a beginner's guide
by Rory McMullan
Gives support and encouragement to get
to work by bike.
£4.95 paperback

HAVE
YOU
TURNED
OFF
YOUR
COMPUTER?